YOUR KNOWLEDGE HAS VALUE

- We will publish your bachelor's and
  master's thesis, essays and papers

- Your own eBook and book -
  sold worldwide in all relevant shops

- Earn money with each sale

Upload your text at www.GRIN.com
and publish for free

Mechanical Robots in Action. A Study on Pick and Place Tasks Using Industrial Automation

Bandar Hezam

Bibliographic information published by the German National Library:

The German National Library lists this publication in the National Bibliography; detailed bibliographic data are available on the Internet at http://dnb.dnb.de.

ISBN: 9783346979988
This book is also available as an ebook.

ASIA PACIFIC UNIVERSITY OF
TECHNOLOGY & INNOVATION

LAB REPORT
INTERMEDIATE ROBOTICS
EE010-3.5-2

TITLE	ABB ROBOT PROGRAMMING FOR PICK & PLACE
NAME	Bandar Naji ALI Hezam
DUE DATE	9 /August /2019

Acknowledgements

In preparation of my report, I had to take the help and guidance of some respected persons, who deserve my deepest gratitude. As the completion of this report gave me much pleasure, I would like to show my gratitude to Dr. ARUN SEERALAN BALAKRISHNAN, Course Instructor ITR, on APU University for giving me a good guideline for report throughout numerous consultations. I would also like to expand my gratitude to all those who have directly and indirectly guided me in writing this Assignment.

In addition, a thank you to Mr. NAZRI HADI, who introduced me to the Methodology of work, and whose passion for the "underlying structures" had lasting effect. Many people, especially my classmates have made valuable comment suggestions on my paper which gave me an inspiration to improve the quality of the report. I would also like to express my gratitude to APU's technicians of the laboratory for giving me the resources needed to complete this assignment.

Abstract

This experiment is for the most part the utilization of the mechanical robots to perform various tasks, and it's to display the outcomes and discoveries of the examination directed by understudies utilizing a modern report to perform pick and place from and to a pallet task. For this reason, a product is utilized to program the robot to achieve the assignment. The report begins with short prologue to furnish the readers with adequate specialized foundation to comprehend the remainder of the report, and after that shows the Apparatus and clarifies the system. Later on, in a different area, the program utilized is clarified in detail, before talking about the most exceptional discoveries of the investigation, which incorporates a few issues that must be tended to for better activity of the robot and puts a few proposals on the table. At last, an end session is to whole up the discoveries, and last session is for the references

Table of Contents

List of Figures

Objective

The main objective of this lab experiment was to program the ABB robot to properly and accurately pick and place the pegs and palletize it on the pallets.

Introduction

Engineering is essentially expressed as the use of arithmetic, science, economy and changing over the hypothetical examinations into handy applications to imagine, create, plan and improve structures, frameworks, programming, machines, materials, segments, and arrangements. The specialization of building is amazingly wide and envelops a scope of increasingly concentrated fields of designing, and each field has a progressively explicit accentuation on territories of connected science, sorts of use and innovation which is in accordance with the outrageous advancement of innovation. One of those sub-fields of building is Mechatronics designing. Mechatronics building is a multidisciplinary field of science which involves a blend of various fields of designing in particular; Mechanical designing, electrical building, media transmission building, control building, electronic, PC designing and framework building.

Apply autonomy is the investigation of robots. Robots are machines that can be utilized to do tasks. A few robots can do work independent from anyone else. Different robots should consistently have an individual guiding them. In this investigation we will utilize The IRB 120 robot is the most recent expansion to ABB's new fourth-age of mechanical innovation as appeared in figure1, to pick and place the pegs into the bed to make the letter 'P' complete .we will utilize RobotStudio programming to program the robot. ABB is a pioneer in power and mechanization innovations that empower utility and industry clients to improve their presentation while bringing down natural effect. The ABB Group of organizations works in around 100 nations and utilizes around 140,000 individuals.

The upsides of the IRB 120 robot can be list in three most significant points of interest, first bit of leeway is Compact and lightweight, it's decreased load of just 25kg. Second is Multipurpose, numerous things can accomplish for models in enterprises including the Inspections of PC sheets, sustenance bundling, sun oriented and therapeutic Ideal for lab computerization and medication fabricating tasks. Third is Fast and coordinated, designed with a light, aluminium structure, the amazing smaller engines guarantee the robot is empowered with a quick increasing speed, and can convey precision and deftness in any application. As we will utilize the robot in our trial in the pick and spot the peg and palletize it on the pallets.by utilizing the Robot Studio to program the robot.

Figure 1:ABB Robot IRB 120

Apparatus

We used some of the equipment in this experiment to achieve our object, as shown in table1

Table 1: list of the apparatus

Equipment	Explanation	picture
- ABB Robot IRB 120	IRB 120 robot is a small robot, fast, agile and easy to use. Used to pick and place.	
- Robot Controller	It is used to move the robot manually and also can program the robot easily.	
- Pegs Pallet	The pegs pallet has five rows and seven columns with a different pegs colors.	
- "VEPRO" pallet	The pallet has five different letters and each letter has some of the holes to place the pegs on it.	
- Ruler	It is a tool used to measure the distances between the pegs and the pallet.	

Programming

Theoretical Review

Right hand rule is important rule to know the x, y and z axis. Z-up typically used by designers as shown in figure2.

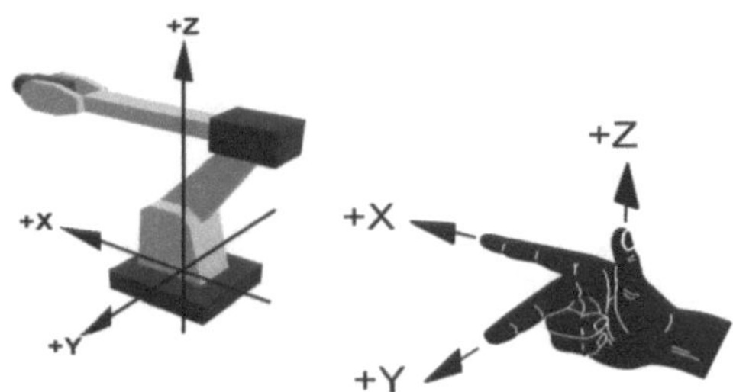

Figure 2:right hand rule

ABB Robot IRB 120, has six axis as shown in figure3. DOF (Degree of freedom) is the number of independent parameters or movement which is needed to uniquely define its position in the free-space at any instant of time.

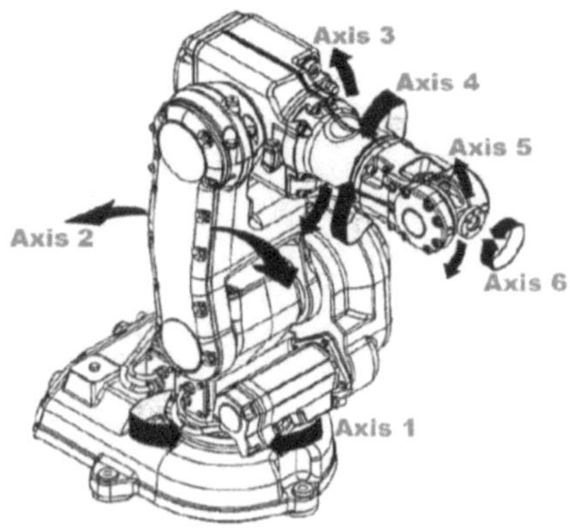

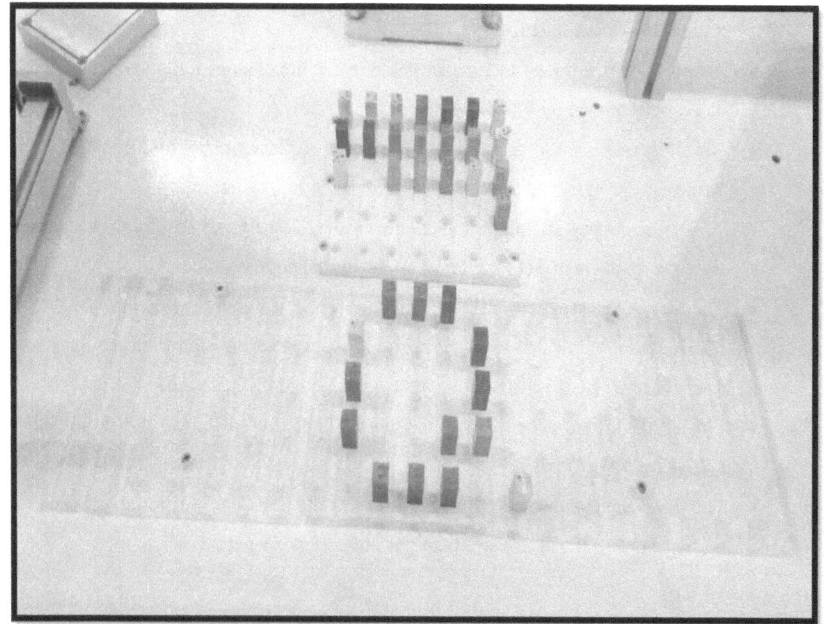

Figure 4:The letter of Q

The letter of Q has been chosen from the list to show the ability of comprehending the programing and the accurse of piking the picas of p20 and place them in p10. The task has accomplished successfully and accurately.

Procedure

- Firstly, measurements are to be taken to identify the distances between the pick points and place points. After that, programming the robot using the flex pendant is begun.
- The origin of the three axes (x, y, z) is initialized by adjusting the robotic arm through the joystick and them modifying the current position, (it's recommended to give it a special name for identification, in our case it's named "pHome"). The code line looks like this:
- Next, using the offset function, the dimensions of the peg point are initialized, so the arm can go to that point.
- Set function is used to grip the peg next, after adjusting it to work with gripper, and the code looks like the following:

5

- The next line of the code is to return the arm to the position prior to gripping (to avoid hitting other pegs mistakenly).
- At this point the offset function is used to set the dimensions for the target point. This could be achieved by following line of code:
- Next, the arm should go closer to the point of placement, which requires only a change in "z" axis; and the code becomes as the following:
- To release the peg and place it in the destined hole, the Reset function is used, which does the opposite of Set function. The line of the code become
- The second last step before finishing with this peg, is to return the arm to the position before coming down to place, to avoid any possible collision with the pegs while going back to get the next peg, and the same line of code is used.
- Finally, the arm is to be returned to the initial position (pHome), to pick up the next peg.
- The dimensions are to be sat properly so that the arm is at position to pick the next peg, and the all the above steps are repeated for every peg.

To make the robot work and make the picking and placing, it was vital to program it. The programming was done in the Robot Studio Software. Right hand rule is important rule to know the x, y and z axis. Z-up typically used by designers as shown in figure1.

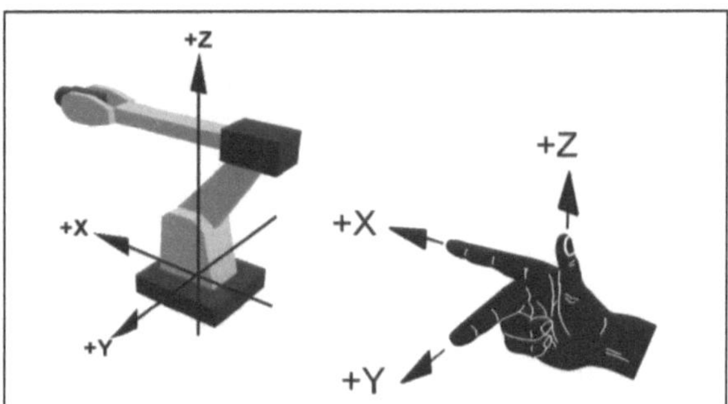

Figure 5:Right Hand Rule

ABB Robot IRB 120, has six axes as shown in figure2. DOF (Degree of freedom) is the number of independent parameters or movement which is needed to uniquely define its position in the free-space at any instant of time.

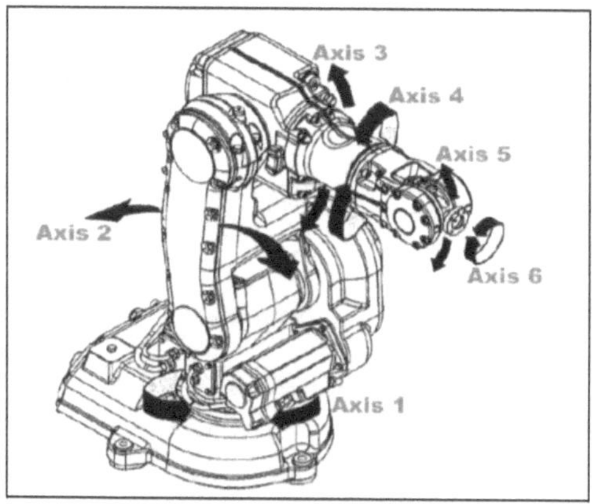

Figure 6:6 ABB

There are 11 pegs that were picked from the picking pallet and placed on the placing pallet to form the letter Q. The first step that was done in the program, was to make the offset points. There were three offset points made. The first offset point, which is p10, has been fixed at the co-ordinates shown in Figure 1. The second offset point, which is p20, has been fixed just above picking pallet. The third offset point, which is p30, has been fixed above the placing pallet. The exact positions of p20 and p30 are shown in Figure 3.

```
1   MODULE MainModule
2     CONST robtarget p10:=[[301.38,-3.37,576.90],[0.0396738,-0.0151978,0.999016,-0.0127212],[-1,-1,-1,0],[9E+09,9E+09,9E+09,9E+09,9E+09,9E+09]];
3     CONST robtarget p20:=[[249.38,-33.47,287.36],[0.0396835,-0.0152107,0.999015,-0.0127433],[-1,-1,-1,0],[9E+09,9E+09,9E+09,9E+09,9E+09,9E+09]];
4     CONST robtarget p30:=[[386.50,23.07,294.43],[0.0396197,-0.0153126,0.999016,-0.0127475],[0,-1,0,0],[9E+09,9E+09,9E+09,9E+09,9E+09,9E+09]];
```

Figure 7:Offset points

After initializing the offset values, the main program starts as shown in Figure 4. Firstly, it will start by using the MoveL function which is command that tells the robot to Move in Linear motion. All commands in Figure 2 are started with a MoveL function. Firstly, it will move in linear motion to the first offset point p10. V1000 indicates the speed that the robot arm has. Fine indicates the zone data of the robot where tool0 indicates that there is no tool. After that, it will move in the same motion towards p20, which is the offset of the picking pallet. Then, it will move -80mm towards the z direction. Then, the function Set was used to activate the gripper so that it grabs the peg. After that, it will move back to the p30 point after going to the p20 point. Then it will go -0.5 mm from the x axis and -82mm to drop the peg and releases it. It will release the peg because of the Rest gripper command.

```
PROC main()
    MoveL p10, v1000, fine, tool0;
    MoveL p20, v1000, fine, tool0;
    MoveL Offs(p20,0,0,-80), v1000, fine, tool0;
    Set DO16_GRIPPER_ON;
    MoveL p20, v1000, fine, tool0;
    MoveL p30, v1000, fine, tool0;
    MoveL Offs(p30,-0.5,0,0), v1000, fine, tool0;
    MoveL Offs(p30,-0.5,0,-82), v1000, fine, tool0;
    Reset DO16_GRIPPER_ON;
```

Figure 8:First Peg

This process was done to pick and place one peg from the picking and placing pallet in order to form the letter Q. The same steps have been executed to move the second peg where firstly, it will move in linear motion towards the offset point p30 with a defined speed, zone data and type of tool. Then, it will move 28 mm from the offset point p20 in y axis. After that, it will pick peg by going -80 mm in the z direction and it will set the gripper on which will pick the peg from the pallet. Then, it will move 28mm from the offset point p20. Subsequently, it will move 29.5 mm and -28 mm from the offset point p30 in the x axis and y axis respectively. Then, from the last position, it will go -82mm to place the peg by the help of the function Reset.

```
MoveL p30, v1000, fine, tool0;
MoveL Offs(p20,0,28,0), v1000, fine, tool0;
MoveL Offs(p20,0,28,-80), v1000, fine, tool0;
Set DO16_GRIPPER_ON;
MoveL Offs(p20,0,28,0), v1000, fine, tool0;
MoveL Offs(p30,29.5,-28,0), v1000, fine, tool0;
MoveL Offs(p30,29,-28,-82), v1000, fine, tool0;
Reset DO16_GRIPPER_ON;
```

Figure 9:Second Peg

After placing the second peg into the pallet, it was the third's peg time to be inserted. This process started with a linear movement from the offset point p30. The movement was towards 29.5 and -28 in x axis and y axis respectively. Then, it will move 55mm in the y axis from the offset point p20. Then, it will go -80 mm in the z-axis from the last point. The gripper will be set to grip the peg from the placing pallet. Then, it will go 54 mm from the y axis from the offset point p20. Then, it will move 58mm and -30mm in the x-axis and the y-axis from the offset point p30. Then, it will go -82mm from the last point. Finally, the gripper will be reset, and the peg will be dropped.

```
MoveL Offs(p30,29.5,-28,0), v1000, fine, tool0;
MoveL Offs(p20,0,55,0), v1000, fine, tool0;
MoveL Offs(p20,0,56,-80), v1000, fine, tool0;
Set DO16_GRIPPER_ON;
MoveL Offs(p20,0,54,0), v1000, fine, tool0;
MoveL Offs(p30,58,-30,0), v1000, fine, tool0;
MoveL Offs(p30,57.5,-29.5,-82), v1000, fine, tool0;
Reset DO16_GRIPPER_ON;
```

Figure 10:Third Peg

The process of placing all other pegs from the picking pallets to the placing pallets are the same. However, the difference is only in the co-ordinates of picking and placing. The rest of codes are placed in the Appendix.

Questions

1- Discuss the importance of safety guidelines in operating an industrial robot.

Safety of human being and the robots is the most important priority undertaken by many manufacturers. Robots are ought to be safe but, anything can happen either due to manufacturer error or human error. Many human beings get injured due to these types of machinery or getting damaged. Therefore, safety guidelines are introduced. This is important where it is indicated in one of the three fundamentals laws of robotics where the first law of robotics is: a robot may not injure a human being or, through inaction, allow a human being to come to harm. The user that will be operating the industrial robot must have adequate knowledge about safety procedures beforehand. Robot can operate in a speed and force that sufficient to injure or kill a person if it is not being operated well. Therefore, introducing safety guidelines are important where it includes the way of operating a robot properly. The guideline is also vital because it tells the user the limitation of the robot. Knowing this, the user will not take the risk to exceed the usage. This will ensure the user understand on how the robot works and they can build safety precaution, at the same time knowing how to stop the machine when something happens. That is why safety guidelines must be first contributed by the manufacturer and must be learnt by the user of any industrial robot. These guidelines ensure the safety working conditions on the human and the robot. One important guideline is having the knowledge from training on how to safely operate the robot. Having enough robot training can help avoid serious injuries as these robots exert thousands of tons of pressure and crush anything on their path.

2- Discuss the importance of the "Emergency Switches" in a robotic work cell.

Emergency switches are being implemented in all the industrial robots. It is built in every robotic work cell. It is to protect the users from injury, or any other worse case. Emergencies switches are vital as it will stop the machine at the earliest time, instead of using the proper way to stop the machine. For example, if an operator met a condition where the machine is moving too fast and might cause danger to the operator and the machine needs to be stopped at the earliest time to prevent any damage. The colleague nearby can help and stop the machine by pressing emergency button before the operator himself can react accordingly. A life might be saved by having the emergency switches. Moreover, the damage of the machine can be reduced. As it was observed, ABB robot has two emergency switches. The first emergency switch is to stop Teach Pendant. While the other one is the guard stop (enabling device). If the robot is moving too fast or out of control, user can press release the motor on and twist the emergency switch. The control panel of the machine has an emergency switch too

3- The IRB 120 ABB robot is a 6 Axis articulated arm robot which you have utilised to run and pick and place operation. Explain the limitation when this articulated arm robot is replacing with SCARA and Cartesian robot for the same operation.

SCARA is known as Selective Compliance Assembly Robot Arm. SCARA moves in X, Y and Z planes, and having a theta axis at the end of Z plane for the end effector to rotate. Cartesian robot has only linear actuators. Hence, it can only make movement in X, Y and Z. SCARA is good for vertical assembly operations. SCARA's arm is a lever, which will limit the reachable area of the arm. When the arm extends, the joints will need to robust bearings and need the help of high-torque motor to handle the load. Cartesian robot has limitation on its movement too. It will only work inside the framework's confines. (Vaughn, 2013) The pick and place of the metal pieces need both linear and rotation movements. SCARA supports rotational movements. Therefore, it is more suitable than Cartesian robot. For pick and place of metal pieces, it does not need much range of movement and power. Hence, both SCARA and six articulated robot arms is suitable. The main difference between these two robot arms is that SCARA has more limitation in movement, while six articulated robot arms can move freely as six joints has its own stepper motor. It can be deduced that cartesian robot is not suitable as it has many limitations while SCARA can be used as long as the pallet is not far away from the robot arm, and the movement for vertical plane is not much as it is not suitable for manipulating objects in vertical plane.

Discussion

The objective of this experiment has been successfully achieved where a program has been developed to pick and place pegs from the pick pallet and to the place pallet by using the ABB robot. The understanding of using the industrial robot have been gained by first simulating and using codes to program the simulated version in the Robot Studio software. Calibration was done to get the desired output. Each peg was adjusted accurately to make it fit onto the pallet which would form a Q letter. However, the desired output was not achieved from the first time. This was due to different reasons and challenges.

One of the main challenges faced was that there was a huge difference between the simulated version of the experiment and the actual robot. Even though that the measurements were taken as accurate as possible but while applying the programmed code, the pegs were not positioned in the right place. This was because of the parallax error that was caused while measuring the distances between each peg. The coordinates had a difference of -/+ 20 mm from the actual placement position. Therefore, to avoid this problem, the placement of the pegs that had this error was done manually. However, there was no other way except to estimate the nearest possible position of those pegs and to try them slowly so that no damages accrue on the robot or the pallets. After adjusting the placement of the pegs manually, the pegs were successfully placed in its pallet.

One more challenge that was faced was the position of the robot arm. The original position of the robot was set to one point. However, if any peg was hit or forced to enter its place, the positioning and the placement of the robot arm changes a bit which causes a change in maintaining the right position of the peg. Similarly, if the robot is to be maintained, the position of the robot changes. When the program was executed, the positioning of all the pegs changed and repositioning was needed in order to make it fit.

All these challenges faced affected the process of getting the desired output. However, these challenges were overcome by making sure that the positioning of the robot arm did not change while making this experiment, and if it did, the positions of the pegs are to set accordingly. One thing that can be done to improve this experiment is to make accurate measurements where the human error must be avoided in order to achieve the wanted measurements.

Conclusion

In conclusion, the objectives and learning outcomes of this experiment have been successfully achieved where a program was made using Robot Studio. The program made was first tested and simulated in the software. Then, it was implemented to the robot arm. However, there were challenges faced during the process like the positioning of the pegs to obtain a letter Q. The most challenging part was getting the accurate distances so that it can be placed in the pallet. But some techniques were used to overcome these challenges where the outcome was achieved. It was also found out that a slight change in the positioning of the robot arm can make a huge difference in the placement of the pegs. Therefore, before executing the experiment, it was needed to make sure of the placements of the pegs so that no damage occurs to the robot arm or the peg.

References

Anon., 2019. https://ifr.org/robot-history.

Anon., 2019. *https://new.abb.com/products/robotics/home/about-us/the-industrial-robot,* s.l.: ABB.

Anon., 2019. *https://www.abb-conversations.com/2013/09/40-years-of-robotic-welding-and-cutting-weve-come-so-far/,* s.l.: s.n.

Anon., 2019. https://www.robotics.org/content-detail.cfm/Industrial-Robotics-News/ABB-Robotics-to-introduce-new-robots-turnkey-manufacturing-cells-and-enhanced-technology-at-IMTS-2014-in-Chicago/content_id/4978. *RAIA*, 9 8.

Appendix

```
MODULE MainModule
      CONST  robtarget  p10:=[[301.38,-3.37,576.90],[0.0396738,-0.0151978,0.999016,-
0.0127212],[-1,-1,-1,0],[9E+09,9E+09,9E+09,9E+09,9E+09,9E+09]];
      CONST robtarget p20:=[[249.38,-33.47,287.36],[0.0396835,-0.0152107,0.999015,-
0.0127433],[-1,-1,-1,0],[9E+09,9E+09,9E+09,9E+09,9E+09,9E+09]];
      CONST  robtarget  p30:=[[386.50,23.07,294.43],[0.0396197,-0.0153126,0.999016,-
0.0127475],[0,-1,0,0],[9E+09,9E+09,9E+09,9E+09,9E+09,9E+09]];
      PROC main()
            MoveL p10, v1000, fine, tool0;
            MoveL p20, v1000, fine, tool0;
            MoveL Offs(p20,0,0,-80), v1000, fine, tool0;
            Set DO16_GRIPPER_ON;
            MoveL p20, v1000, fine, tool0;
            MoveL p30, v1000, fine, tool0;
            MoveL Offs(p30,-0.5,0,0), v1000, fine, tool0;
            MoveL Offs(p30,-0.5,0,-82), v1000, fine, tool0;
            Reset DO16_GRIPPER_ON;
            MoveL p30, v1000, fine, tool0;
            MoveL Offs(p20,0,28,0), v1000, fine, tool0;
            MoveL Offs(p20,0,28,-80), v1000, fine, tool0;
            Set DO16_GRIPPER_ON;
            MoveL Offs(p20,0,28,0), v1000, fine, tool0;
            MoveL Offs(p30,29.5,-28,0), v1000, fine, tool0;
            MoveL Offs(p30,29,-28,-82), v1000, fine, tool0;
            Reset DO16_GRIPPER_ON;
            MoveL Offs(p30,29.5,-28,0), v1000, fine, tool0;
        MoveL Offs(p20,0,55,0), v1000, fine, tool0;
        MoveL Offs(p20,0,56,-80), v1000, fine, tool0;
        Set DO16_GRIPPER_ON;
        MoveL Offs(p20,0,54,0), v1000, fine, tool0;
        MoveL Offs(p30,58,-30,0), v1000, fine, tool0;
        MoveL Offs(p30,57.5,-29.5,-82), v1000, fine, tool0;
        Reset DO16_GRIPPER_ON;
        MoveL Offs(p30,58,-29.5,0), v1000, fine, tool0;
```

```
MoveL Offs(p20,0,85,0), v1000, fine, tool0;
MoveL Offs(p20,0,85,-80), v1000, fine, tool0;
Set DO16_GRIPPER_ON;
MoveL Offs(p20,0,82,0), v1000, fine, tool0;
MoveL Offs(p30,86.5,-3.5,0), v1000, fine, tool0;
MoveL Offs(p30,87,-2.5,-80), v1000, fine, tool0;
Reset DO16_GRIPPER_ON;
MoveL Offs(p30,88.5,-2.5,0), v1000, fine, tool0;
MoveL Offs(p20,-1.5,113,0), v1000, fine, tool0;
MoveL Offs(p20,-1.5,114,-80), v1000, fine, tool0;
Set DO16_GRIPPER_ON;
MoveL Offs(p20,0,110,0), v1000, fine, tool0;
MoveL Offs(p30,86.5,24.5,0), v1000, fine, tool0;
MoveL Offs(p30,86.5,24.5,-80), v1000, fine, tool0;
Reset DO16_GRIPPER_ON;
MoveL Offs(p30,88.5,24.5,0), v1000, fine, tool0;
MoveL Offs(p20,-2,138,0), v1000, fine, tool0;
MoveL Offs(p20,-2,142.5,-80), v1000, fine, tool0;
Set DO16_GRIPPER_ON;
MoveL Offs(p20,0,138,0), v1000, fine, tool0;
MoveL Offs(p30,59,51.7,0), v1000, fine, tool0;
MoveL Offs(p30,57.5,51.7,-80), v1000, fine, tool0;
Reset DO16_GRIPPER_ON;
MoveL Offs(p30,59,51.7,0), v1000, fine, tool0;
MoveL Offs(p20,-4,170,0), v1000, fine, tool0;
MoveL Offs(p20,-4,170,-80), v1000, fine, tool0;
Set DO16_GRIPPER_ON;
MoveL Offs(p20,0,170,0), v1000, fine, tool0;
MoveL Offs(p30,28.5,52.5,0), v1000, fine, tool0;
MoveL Offs(p30,28.5,52.5,-80), v1000, fine, tool0;
Reset DO16_GRIPPER_ON;
MoveL Offs(p30,32.5,51.7,0), v1000, fine, tool0;
MoveL Offs(p20,26,170,0), v1000, fine, tool0;
MoveL Offs(p20,26,170,-80), v1000, fine, tool0;
Set DO16_GRIPPER_ON;
```

```
        MoveL Offs(p20,29,170,0), v1000, fine, tool0;
        MoveL Offs(p30,-1,28,0), v1000, fine, tool0;
        MoveL Offs(p30,-1,28,-81), v1000, fine, tool0;
        Reset DO16_GRIPPER_ON;
        MoveL Offs(p30,2,28,0), v1000, fine, tool0;
        MoveL Offs(p20,27.5,145,0), v1000, fine, tool0;
        MoveL Offs(p20,27.5,145,-80), v1000, fine, tool0;
        Set DO16_GRIPPER_ON;
        MoveL Offs(p20,29,145,0), v1000, fine, tool0;
        MoveL Offs(p30,57,30,0), v1000, fine, tool0;
        MoveL Offs(p30,57,29.5,-81), v1000, fine, tool0;
        Reset DO16_GRIPPER_ON;
        MoveL Offs(p30,58,30,0), v1000, fine, tool0;
        MoveL Offs(p20,25,117,0), v1000, fine, tool0;
        MoveL Offs(p20,25,117,-80), v1000, fine, tool0;
        Set DO16_GRIPPER_ON;
        MoveL Offs(p20,29,117,0), v1000, fine, tool0;
        MoveL Offs(p30,85,55,0), v1000, fine, tool0;
        MoveL Offs(p30,85,55,-80), v1000, fine, tool0;
        Reset DO16_GRIPPER_ON;
        MoveL Offs(p30,90,55,0), v1000, fine, tool0;
        MoveL Offs(p20,26,90,0), v1000, fine, tool0;
        MoveL Offs(p20,26,90,-80), v1000, fine, tool0;
        Set DO16_GRIPPER_ON;
        MoveL Offs(p20,29,90,0), v1000, fine, tool0;
        MoveL Offs(p30,113.5,82.5,0), v1000, fine, tool0;
        MoveL Offs(p30,113.5,82.5,-81), v1000, fine, tool0;
        Reset DO16_GRIPPER_ON;
        MoveL Offs(p30,119,81,0), v1000, fine, tool0;


ENDPROC
ENDMODULE
```

YOUR KNOWLEDGE HAS VALUE

- We will publish your bachelor's and
  master's thesis, essays and papers

- Your own eBook and book -
  sold worldwide in all relevant shops

- Earn money with each sale

Upload your text at www.GRIN.com
and publish for free